Math Puzzles and Logic Problems: Book Two

A Cipher and Sage Large Print Edition

Name _____

Phone _____

Introduction

The first Math Puzzles and Logic Problems Book contained a variety of puzzles that I was familiar with solving. As I continue researching types of math puzzles availabe, I find interesting and new types to work. This book follows the same format as the first one.

The first chapter contains two of each type of puzzle in the book - one with instructions and one without. Both are easier versions of the puzzles to allow the solver to familiarize themself with the mechanics and rules of the challenge.

The remainder of the book contains two of each type of puzzle in each chapter. Chapters are arranged by difficulty with the idea that the puzzles will become more difficult to solve as the book goes on.

We kept some of the puzzle types of book one (Logic grids, Kakuro, Sudoku, and Number Search and introduce three new puzzles. In subsequent volumes, we will continue mix old with new puzzle types.

The last chapter of this book contains solutions to all of the puzzles. I hope you enjoy this book of math puzzles as much as I enjoyed creating it. Happy solving!

Chapter 1
Easy

Number Search

Instructions - Find the numbers in the list below the puzzle. Numbers can run forward or backward, up or down, diagonal forward or diagonal backward.

```
9  1  0  3  8  0  3  9  7  4
7  9  9  7  3  2  6  3  4  6
5  0  1  3  5  1  1  0  2  9
3  1  1  9  2  3  1  1  2  9
6  0  1  5  6  1  7  5  8  0
9  0  9  5  5  9  8  1  3  2
7  5  3  5  1  8  5  6  8  4
9  1  4  5  8  8  3  9  4  3
3  6  5  5  7  4  9  4  6  5
7  5  3  9  5  0  7  8  8  4
```

- ☐ 113
- ☐ 909
- ☐ 615
- ☐ 5115
- ☐ 2011
- ☐ 557
- ☐ 559
- ☐ 7357
- ☐ 5617
- ☐ 8839

Number Search 2

```
2  3  1  6  1  0  6  0  3  4
9  2  5  8  9  1  2  9  2  9
6  1  3  2  0  7  3  0  4  0
8  0  1  0  6  8  3  9  3  7
9  6  0  9  7  1  7  0  5  5
3  8  7  3  3  5  4  3  8  4
3  4  9  4  7  7  6  4  6  7
0  7  8  5  9  3  6  2  4  4
1  8  7  9  3  0  7  5  4  6
4  0  3  0  0  3  5  1  2  5
```

☐ 664   ☐ 8157

☐ 0798   ☐ 7379

☐ 8037   ☐ 8534

☐ 230   ☐ 893

☐ 2921   ☐ 600

Kakuro

In a Kakuro puzzle, you are given the solution in the grey squares for the set of white squares that the arrow points toward. You may only use the numbers 1 through 9 to solve and these numbers may only be added together to get the total.

Hint: Here are some common combinations - 1+2+3 = 6, 1+2+4 = 7, 1+2+3+4 = 10, 3 can only be 1+2 and 4 can only be 1+3, 16 can only be 9+7 and 17 can only be 9+8.

Kakuro 2

	14↓	29↓	22↓	15↓	17↓	14↓	6↓	
36→								■
35→							12↓	■
15→				4↓	8→ 29↓			29↓
23→						3→		
■	5→ 12↓			2→		10→ 7↓		
3→			13↓	17→ 6↓				
32→							8→ 2↓	
2→		21→				11→		

Sudoku

Sudoku also only allows the usage of the numbers 1 through 9 and only allows them once. In this case, 1-9 can only be found once in each square, in each row and in each column. Some solvers like to write the possibilities up in the corners and then cross them out until there is only one left.

3			7	6	1	9		4
1	6	7	4	8	9			
4				5	3	1		
			1			4		5
2	4		5			8		9
7	9		3				2	1
5	7		6	1	4			3
6	1	3				5		8
	2		8	3			1	6

Sudoku 2

	7			9				
5		3	7	4		1	9	
8		6		3				
7	5	8	4	2	9			
			3	7	8	4	2	
3		2	5	6				9
9		1	6		7	2	5	
2	6			5		8		1
4			2	1		9		6

Logic Puzzle

Logic grids are story based puzzles. You receive bits and pieces of information about a set of related things, places, or people. From these bits and pieces, you must deduce the specific relationships. The grid is merely a helper in solving the puzzle and some people prefer other methods of solving. Use whatever works for you.

		Meal					Cars				Event				
		Lasagna	Chicken	Sushi	Salad	Apple Pie	Honda	Toyota	Ford	Tesla	Nissan	Promotion	Graduation	Engagement	House	New Baby
Friends	Allan															
	Brenda															
	Craig															
	Donna															
	Ella															
Event	Promotion															
	Graduation															
	Engagement															
	House															
	New Baby															
Cars	Honda															
	Toyota															
	Ford															
	Tesla															
	Nissan															

The Friends' Gathering

Five friends – Allan, Brenda, Craig, Donna, and Ella – decide to meet for a meal. Each brings a different dish: lasagna, fried chicken, sushi, salad, and apple pie. Each of them is celebrating a different recent life event: a promotion, a graduation, an engagement, a house purchase, and the birth of a new child. They also arrive at the gathering in five different cars: a Honda Civic, a Toyota Corolla, a Ford Mustang, a Tesla Model 3, and a Nissan Altima.

Clues:

1. Allan, who didn't bring sushi, recently welcomed a new child.
2. Brenda arrived in the Toyota but didn't bring lasagna.
3. The person who celebrated an engagement brought fried chicken.
4. The friend who arrived in the Honda Civic brought sushi.
5. Donna celebrated her new house and did not arrive in the Nissan or the Ford Mustang.
6. Craig did not recently graduate or drive a Ford Mustang or bring Sushi to the party.
7. Ella's left hand flashed when she arrived and opened the door of her Tesla.
8. Neither Brenda nor Allan brought Salad.

Logic Puzzle 2

		Cabin Color					Activity					Beverage				
		Red	Blue	Green	Yellow	White	Hiking	Reading	Swimming	Bird Watch	Painting	Tea	Coffee	Lemonade	Hot Choco	Water
Friends	Marco															
	Nina															
	Oliver															
	Penny															
	Quinn															
Beverage	Tea															
	Coffee															
	Lemonade															
	Hot Choco															
	Water															
Activity	Hiking															
	Reading															
	Swimming															
	Bird Watch															
	Painting															

The Weekend Retreat

Five friends – Marco, Nina, Oliver, Penny, and Quinn – went on a weekend retreat. Each stayed in a different colored cabin: red, blue, green, yellow, and white. They each enjoyed a different activity: hiking, reading, swimming, bird-watching, and painting. Additionally, each friend has a favorite beverage: tea, coffee, lemonade, hot chocolate, and water.

Clues:

1. Marco stayed in the red cabin and did not go swimming.
2. The person in the green cabin enjoys bird-watching.
3. Quinn drank tea but didn't stay in the white, yellow, or green cabin.
4. Oliver drank lemonade and did not stay in the yellow or green cabin.
5. Penny enjoys swimming but did not drink coffee or hot chocolate.
6. The person in the green cabin drank hot chocolate.
7. The person who enjoyed hiking drank coffee.
8. The person who drank lemonade enjoyed reading in his white cabin.

Calcudoku

Each puzzle consists of a grid containing blocks surrounded by bold lines. Fill the empty squares so each number appears appear exactly once in each row and column and the numbers produce the result shown in the top left corner of the dark outlined bloc according to the math operation shown. In a 4x4 grid, the numbers available are 1, 2, 3, 4. In a 5x5 grid, the numbers are 1,2, 3, 4, 5 and so on.

7+			3
		3	7+
		6+	
8+	6+		

Calcudoku 2

4	6+		
3+			
5+	3	7+	
	5+		7+

Number Place

Number Place is similar to a crossword puzzle in that each number in the clues section has to be placed either across or down as indicated. The numbers will only fit in the puzzle one way. Once completed, all of the open squares will be filled. Squares with no numbers have already been blacked out.

ACROSS

222, 1223

DOWN

2123, 22, 122

Number Place 2

Hitori

Every grid comes with a number in all squares.

The object is to shade squares so:

- No number appears in a row or column more than once.

- Shaded squares do not touch each other vertically or horizontally.

- When completed, all un-shaded squares create a single continuous area.

1	2	2	4
1	4	1	3
3	2	4	2
4	3	2	2

Hitori 2

4	1	1	4
2	4	1	3
3	3	4	2
4	2	3	1

Chapter 2
Medium

Hitori 3

2	5	4	2	3
1	3	3	2	5
4	1	5	2	2
5	4	2	4	1
3	2	1	5	4

Number Place 3

ACROSS

42, 44121, 4112

DOWN

344, 11, 14, 44, 22

25

Logic Puzzle 3

		Pets					House Color					Fruits				
		Cat	Dog	Rabbit	Parrot	Fish	Brown	Pink	Green	Blue	White	Apple	Banana	Cherry	Date	Elderberry
Neighbors	Sarah															
	Tim															
	Ursula															
	Victor															
	Wendy															
Fruits	Apple															
	Banana															
	Cherry															
	Date															
	Elderberry															
House Color	Brown															
	Pink															
	Green															
	Blue															
	White															

The Pet Owners

Five neighbors – Sarah, Tim, Ursula, Victor, and Wendy – each own a different pet: a cat, a dog, a rabbit, a parrot, and a fish. They each live in a house with a different color: brown, pink, green, blue, and white. They also have different favorite fruits: apple, banana, cherry, date, and elderberry.

Clues:

1. Sarah lives in the white house and doesn't own the parrot.
2. The cat owner lives in the green house.
3. Tim's favorite fruit is apples, but he doesn't live in the green or blue house.
4. Ursula lives in the blue house but doesn't own the rabbit or the fish.
5. The person who lives in the brown house owns the rabbit.
6. Wendy loves banana and doesn't own the dog.
7. The dog owner loves elderberries.
8. Victor doesn't live in the pink or brown house and his favorite fruit isn't cherry.
9. The fish owner loves apples.

Number Search 3

```
8 7 8 1 6 0 1 0 0 5 3 9
0 8 5 4 6 7 3 3 9 4 1 2
4 7 5 7 3 5 5 5 5 2 4 0
8 9 2 7 1 5 7 8 6 8 7 0
7 0 9 4 8 9 2 8 2 2 2 6
2 9 4 5 0 3 0 6 3 1 6 6
8 5 3 5 4 3 9 9 8 4 5 2
6 0 8 6 4 7 1 3 8 7 9 2
2 9 6 7 0 5 9 5 0 1 6 7
8 5 3 9 0 0 4 5 1 2 4 9
```

- [ ] 80487
- [ ] 72091
- [ ] 7265
- [ ] 832
- [ ] 948
- [ ] 1366
- [ ] 305
- [ ] 2555
- [ ] 82147
- [ ] 0639
- [ ] 9337
- [ ] 8552

Sudoku 3

				2		8		1
	5				9			
		2						
	7			6			5	9
3	9		8	5			1	
				4	2		8	7
5	4			8	6	9		3
6				1				8
7		3			5		2	

Calcudoku 3

6+		5+	
9+		2	8+
		1	
		9+	

Kakuro 3

	38 ↓	3 ↓	34 ↓	30 ↓	10 ↓	14 ↓	6 ↓	7 ↓
37 →								
6 →		19 →					5 → 20 ↓	
8 →		24 → 29 ↓						21 ↓
27 →					14 → 19 ↓			
20 →						10 → 6 ↓		
10 →				23 → 11 ↓				
26 →						■	1 → 5 ↓	
5 →			13 →			5 →		■

Number Search 4

```
6  5  5  0  1  2  8  2  6  3  6  4
7  4  3  8  9  2  5  5  3  2  8  4
6  6  1  4  7  8  6  9  2  8  2  9
6  5  5  6  4  0  3  4  9  2  0  0
3  5  1  1  2  9  0  3  1  3  6  5
6  7  9  0  3  6  7  6  7  5  7  5
2  3  1  4  9  2  9  5  1  0  8  2
3  1  0  0  2  6  4  0  4  1  0  2
6  5  2  4  2  2  8  3  8  9  5  5
5  7  0  7  0  5  7  1  5  1  4  4
```

☐ 9952   ☐ 15135

☐ 00733   ☐ 6200

☐ 7663   ☐ 60078

☐ 6820   ☐ 546

☐ 621   ☐ 57944

☐ 65573   ☐ 849

Number Place 4

ACROSS

22, 34, 2412, 12

DOWN

23, 222, 42, 4232

Logic Puzzle 4

		Flowers					Tool					Season				
		Roses	Tulips	Lilies	Daisies	Sunflowers	Spade	Rake	Hoe	Trowel	Shears	Spring	Summer	Fall	Winter	Rainy
Neighbors	Zack															
	Yara															
	Xavier															
	Yvonne															
	Zoe															
Season	Spring															
	Summer															
	Fall															
	Winter															
	Rainy															
Tool	Spade															
	Rake															
	Hoe															
	Trowel															
	Shears															

Gardening Enthusiasts

Five neighbors - Zack, Yara, Xavier, Yvonne, and Zoe - each have a favorite flower: roses, tulips, lilies, daisies, and sunflowers. They each have a preferred gardening tool: spade, rake, hoe, trowel, and shears. Also, they each have a favorite season: spring, summer, fall, winter, and rainy.

Clues:

1. Zack loves sunflowers but doesn't use the trowel.
2. The one who favors roses always gardens in the summer.
3. Yara uses a rake and her favorite season isn't winter.
4. The person who loves tulips uses a hoe and prefers gardening in the rainy season.
5. Zoe's favorite season is spring but she doesn't like daisies or tulips.
6. Xavier loves the rain and the spring gardener uses a trowel.
7. The gardener who prefers winter uses the shears.
8. Yvonne loves summer and her favorite flower isn't sunflowers or daisies.

Hitori 4

1	2	4	5	3
5	3	1	3	2
2	5	3	1	1
4	1	5	3	5
3	4	5	1	1

Calcudoku 4

4		6+	3+	
7+				7+
		7+	6+	
	14+			8+
	4+		4	5

Sudoku 4

7			2	9		6		
			7			2		
	9	5	4	6	8			
1	8	9	6				7	
6					7		8	2
		2		5		4		
8		1	5	7				9
			1			7		4
	3	7			6	8		

Kakuro 4

Chapter 3
Challenging

Logic Puzzle 5

		Band					Shirt Color					Snack				
		RedValley	SonicBurst	MelodyMinds	LunarTunes	QuantumBeat	Black	White	Blue	Green	Yellow	Popcorn	Nachos	Ice Cream	Pretzel	Fruit Salad
Friends	Amy															
	Brian															
	Carla															
	Derek															
	Evan															
Snack	Popcorn															
	Nachos															
	Ice Cream															
	Pretzel															
	Fruit Salad															
Shirt Color	Black															
	White															
	Blue															
	Green															
	Yellow															

Music Festival Friends

Five friends – Amy, Brian, Carla, Derek, and Evan – attended a music festival. They each saw a different band perform: RedValley, SonicBurst, MelodyMinds, LunarTunes, and QuantumBeats. They each wore a different colored shirt: black, white, blue, green, and yellow. Each also had a different snack at the festival: popcorn, nachos, ice cream, pretzel, and fruit salad.

Clues:

1. Amy wore a white shirt and didn't see SonicBurst or eat nachos.
2. Brian saw QuantumBeats and didn't have ice cream or wear green.
3. The person in the green shirt saw RedValley perform while eating popcorn.
4. Derek had a pretzel but didn't wear a green or blue shirt when he saw MelodyMinds.
5. The person who saw LunarTunes had ice cream.
6. Evan wore a blue shirt and did not eat nachos.
7. The person in the yellow shirt had a pretzel.

Sudoku 5

						1		
	9				1	4		5
			7		8	2		
	5	7	2			8	6	
				8				
4			9					
1		3	8	7				
8	2			3	5			
9			1		2			

Number Search 5

```
9 3 8 2 1 2 8 1 3 6 0 0 2 2 3 6
7 7 9 2 1 8 2 1 9 2 5 8 6 8 1 1
7 7 6 8 2 7 2 8 5 0 8 5 8 6 6 9
3 1 4 0 7 2 1 4 3 9 9 4 5 9 7 7
1 8 5 4 5 5 6 4 6 0 2 3 9 3 1 4
0 5 9 7 3 7 7 1 9 5 3 7 8 8 9 2
7 0 9 8 4 9 5 5 7 4 8 1 6 2 8 5
6 7 1 1 5 4 8 8 5 0 2 4 2 0 7 9
6 9 4 3 4 0 7 9 4 4 4 7 0 4 4 0
4 7 2 9 0 3 5 6 3 9 8 4 6 7 4 1
3 7 7 9 8 5 9 7 4 4 4 6 6 4 9 3
3 7 6 6 6 4 4 3 7 2 1 4 9 2 8 5
5 6 3 8 1 0 9 4 9 5 8 6 7 6 3 8
2 4 0 7 9 1 9 0 7 5 1 3 4 8 3 6
1 4 0 6 5 3 9 2 3 2 7 6 9 2 0 6
6 9 2 3 3 2 6 4 0 0 6 9 9 1 8 1
```

- [ ] 5755
- [ ] 849447
- [ ] 821925
- [ ] 094894
- [ ] 26895
- [ ] 71574
- [ ] 2484
- [ ] 4283

- [ ] 5875
- [ ] 6929
- [ ] 446777
- [ ] 6459
- [ ] 11275
- [ ] 85437
- [ ] 83031

Hitori 5

3	3	5	2	2	1
6	1	5	5	2	3
1	6	2	4	3	5
4	2	5	6	2	4
2	5	1	3	4	1
1	4	6	3	5	2

Calcudoku 5

		3		1-
	14+	13+		
	1		4	
10+	14+	4-		1-

Kakuro 5

	27↓	19↓	■	■	7↓	16↓	45↓	■	14↓	28↓
11→			■	12→ 29↓				16→ 16↓		
6→			29→ 37↓							
30→					16→				1→ 21↓	
18→					24→ 26↓					
■	24→ 21↓						10→ 9↓			
3→		11→ 22↓				25→				
29→						14→ 10↓				14↓
12→				24→ 7↓					2→	
22→					6→ 9↓			1↓	4→	
■	5→		15→			■	1→		8→	

Number Place 5

ACROSS

12, 2335, 11, 14, 55554, 121234

DOWN

31, 52, 112425, 14, 213

Kakuro 6

	7↓	33↓	19↓		19↓	5↓	10↓			45↓
18→			15→ 33↓						1→ 16↓	
	18→ 30↓					1→ 39↓		14→ 4↓		
35→							9→ 16↓			
40→								12→ 13↓		
24→					10→ 21↓				9→ 9↓	
9→			34→ 19↓							
1→		13→ 15↓					11→ 2↓			
39→									4→ 6↓	
12→				6→ 1↓					8→ 7↓	
	18→						8→			

Hitori 6

5	7	4	1	3	6	1
3	5	1	6	2	2	7
3	6	2	5	1	7	7
7	2	5	1	6	6	4
5	1	1	4	3	5	3
6	5	3	7	4	1	2
1	2	7	5	4	4	6

Logic Puzzle 6

		Book					Bag					Beverage				
		Mysteries Of	Ancient Civ	Ocean's Sec	Magic & Myth	Digital Fut	Backpack	Tote	Sling	Messenger	Handbag	Tea	Coffee	Soda	Hot Choco	Water
Friends	Fiona															
	Greg															
	Helen															
	Ian															
	Julie															
Beverage	Tea															
	Coffee															
	Soda															
	Hot Choco															
	Water															
Bag	Backpack															
	Tote															
	Sling															
	Messenger															
	Handbag															

Library Lovers

Five friends – Fiona, Greg, Helen, Ian, and Julie – each borrowed a different book from the library: "Mysteries of Space", "Ancient Civilizations", "Ocean's Secrets", "Magic & Myth", and "Digital Future". They each brought a different type of bag: backpack, tote bag, sling bag, messenger bag, and handbag. Each also enjoyed a different beverage while reading: coffee, tea, soda, water, and hot chocolate.

Clues:

1. Fiona carried a a messenger bag and didn't borrow "Mysteries of Space" or drink soda.
2. Greg read "Digital Future" and his drink wasn't water.
3. The one with the backback borrowed "Magic & Myth".
4. Helen drank coffee but didn't carry a sling bag or a handbag.
5. The librarian noticed a hot chocolate stain on "Ancient Civilizations" and was glad that the person who checked out "Ocean's Secrets" always drank water while she read.
6. Ian didn't carry the handbag and he drank hot chocolate.
7. Julie didn't borrow "Ocean's Secrets" and didn't drink soda.
8. The person with the tote drank soda.

Number Place 6

ACROSS

342126, 6241, 33, 113, 652, 56125, 361, 61

DOWN

16, 46, 23136, 13, 645, 432, 2431, 5316, 216, 511

54

Calcudoku 6

1-		1-		5
	10+		1-	
		2	3-	
2-		4+		
	4		2-	10+

Number Search 6

```
5 7 0 2 7 1 2 5 3 8 9 8 1 4 3 4
9 3 7 9 6 8 6 1 5 8 0 3 6 0 5 4
8 1 0 8 4 0 3 8 8 9 2 4 5 3 9 4
5 6 3 1 9 9 5 3 9 6 4 2 7 9 6 6
9 9 6 4 5 7 5 0 2 6 0 7 1 4 5 3
5 1 0 3 8 0 6 0 6 4 9 5 2 5 8 6
8 5 9 0 5 5 7 2 5 4 1 3 6 3 4 3
9 2 5 8 9 0 4 9 8 1 8 9 0 9 3 5
8 6 0 7 9 6 7 0 3 2 9 3 5 4 1 1
1 6 2 2 5 6 2 9 9 1 5 9 9 0 1 9
9 5 4 0 9 7 5 6 8 4 0 4 0 1 1 0
6 8 4 5 3 7 0 3 1 7 9 7 3 4 8 0
0 8 3 7 8 5 9 8 3 2 1 4 7 1 2 6
1 8 3 9 8 9 9 5 3 0 5 8 9 4 0 0
2 6 2 3 2 8 5 8 7 5 3 7 7 9 5 0
2 6 7 7 5 7 0 7 6 1 5 3 5 9 6 6
```

- [ ] 8474
- [ ] 059037
- [ ] 905981
- [ ] 2811
- [ ] 1084
- [ ] 0381
- [ ] 8321
- [ ] 5377
- [ ] 3589
- [ ] 37031
- [ ] 23076
- [ ] 50278
- [ ] 53058
- [ ] 476553
- [ ] 94141

Sudoku 6

	1		9	8	2			4
2			3	5	4		6	
			8		5			
7				1	9	4		3
			4	6		8		9
8			7		3			
	3	9	6			7		8
		2						6

Chapter 4
Expert

Calcudoku 7

				12x
		5x	8x	
150x	2	4	3x	
12x		30x		10x
4				

Hitori 7

8	6	4	1	6	3	1	1
6	4	1	7	8	2	3	5
5	1	4	4	3	3	8	2
4	1	8	6	7	1	6	6
2	7	3	6	8	8	1	4
5	6	6	8	1	1	4	2
7	8	6	1	3	5	2	3
1	8	7	3	2	4	6	8

Logic Puzzle 7

		Exhibit					Transport					Souvenir				
		Impressions	Sculpture	Abstract	Realistic	Digital	Bus	Bike	Car	Subway	Walk	Poster	Mug	Keychain	Scarf	Notebook
Friends	Lauren															
	Mark															
	Nancy															
	Oscar															
	Penny															
Souvenir	Poster															
	Mug															
	Keychain															
	Scarf															
	Notebook															
Transport	Bus															
	Bike															
	Car															
	Subway															
	Walk															

Art Exhibit Excursion

Five friends – Lauren, Mark, Nancy, Oscar, and Penny – each visited a different art exhibit at the museum: "Impressionist Impressions", "Sculptural Wonders", "Abstract Adventures", "Realistic Renderings", and "Digital Dreams". They each took a different form of transportation to get to the museum: bus, bike, car, subway, and walking. After their visit, they each bought a different souvenir: poster, mug, keychain, scarf, and notebook

Clues:

1. Lauren went to the "Abstract Adventures" exhibit and didn't take the subway.
2. Mark took a bus and didn't buy a keychain or a notebook.
3. The person who bought a keychain with a small version of a sculpture recently bought a car they brought to the museum.
4. The one who visited "Digital Dreams" walked.
5. Nancy didn't visit "Realistic Renderings" or "Impressionist Impressions".
6. Oscar didn't take the car and he bought a poster.
7. Penny bought a mug and didn't visit "Realistic Renderings" or take the subway.
8. The person who saw "Impressionist Impressions" did not buy a scarf or a mug.

Sudoku 7

				2		9	4	
		4						8
5	9				1		6	
			9				1	4
3		8		5	4			
				6				
	3		7		5			9
		9			8		5	6
1			6					

Number Search 7

```
6 4 4 0 5 3 0 5 9 5 0 4 0 0 9 8 3 0 7 1
5 3 2 2 6 1 5 5 5 5 5 0 1 9 4 7 4 5 6 7
6 7 1 1 4 6 3 7 1 6 4 0 4 5 9 3 2 0 5 4
9 5 1 6 2 9 0 9 2 9 1 9 3 3 8 9 7 8 1 1
2 8 1 6 3 9 1 2 6 4 2 2 2 8 7 6 5 2 8 6
7 4 2 0 6 0 0 1 7 5 1 2 3 1 5 6 9 8 9 8
6 8 7 4 5 0 0 7 5 0 0 8 1 6 5 1 8 2 0 9
8 4 4 2 8 7 5 9 7 0 9 9 8 6 8 3 8 1 1 0
6 2 0 2 4 5 2 3 0 9 0 3 8 3 5 2 0 5 3 4
1 7 3 6 0 4 8 1 7 8 1 9 8 5 9 6 6 1 0 9
8 4 8 5 9 0 4 1 2 6 4 7 7 9 2 4 5 6 2 3
7 7 3 9 3 9 0 4 4 0 5 8 9 0 9 3 7 2 0 2
1 9 1 3 8 7 8 7 2 7 6 1 2 3 6 5 3 3 9 6
6 7 9 4 9 4 9 9 9 0 6 1 9 2 2 4 9 4 6 0
4 0 8 7 1 2 8 0 4 0 5 8 9 1 5 7 8 8 2 5
1 4 9 6 9 7 1 2 5 2 8 0 9 1 6 3 4 6 4 5
3 5 6 3 7 8 4 1 0 3 2 6 2 7 7 0 3 1 8 8
8 6 3 2 2 5 2 7 3 7 7 8 9 3 8 7 5 0 4 8
3 2 7 2 2 6 8 3 2 1 8 0 7 9 0 1 1 0 3 2
9 1 7 2 9 4 6 2 4 5 4 9 2 9 5 5 5 0 6 6
```

- [ ] 650985
- [ ] 9983907
- [ ] 14174
- [ ] 00437
- [ ] 12386
- [ ] 7726230
- [ ] 742060
- [ ] 031098
- [ ] 6789848
- [ ] 7000248
- [ ] 45007
- [ ] 67418
- [ ] 739843
- [ ] 314617
- [ ] 7197709
- [ ] 08252
- [ ] 192216
- [ ] 615128
- [ ] 549215
- [ ] 909261

Kakuro 7

	38↓	20↓	36↓	35↓	■	■	12↓	13↓	5↓	10↓
20→					25↓	16→ 45↓				
42→									4→ 30↓	
29→							8→			7↓
9→		19→ 23↓					10↓	15→		
37→								5→ 14↓		31↓
19→					28→ 18↓					
36→							16→ 23↓			
10→			13↓	20→ 5↓				11→		
■	35→ 7↓							8→ 2↓		
19→					12→				4→	

Number Place 7

ACROSS

11, 42, 77, 41, 563, 62125, 426117, 14, 57717, 35, 773, 62, 24, 262

DOWN

164, 62421, 1222517, 241, 22, 15, 755, 154, 33, 472, 13, 6677

Logic Puzzle 8

		Pets					House Color					Pet Food				
		Cat	Dog	Rabbit	Parrot	Fish	Red	Blue	Green	Yellow	White	Whisk&Purr	BarkyMeal	FeatherDelig	Hop&Nibble	AquaFresh
Friends	Alice															
	Bob															
	Claire															
	David															
	Emma															
Pet Food	Whisk&Purr															
	BarkyMeal															
	FeatherDelig															
	Hop&Nibble															
	AquaFresh															
House Color	Red															
	Blue															
	Green															
	Yellow															
	White															

Pet Parade

Five friends — Alice, Bob, Claire, David, and Emma — each have a different pet: cat, dog, parrot, rabbit, and fish. They live on the same street and each of their houses is painted a different color: red, blue, green, yellow, and white. Every morning, they buy a different brand of food for their pets: Whisk & Purr, BarkyMeal, FeatherDelight, Hop & Nibble, and AquaFresh.

Clues:

1. Alice lives in the yellow house and doesn't have a parrot.
2. Bob has a fish and his house isn't yellow or blue.
3. The green house's owner buys BarkyMeal.
4. Claire bought FeatherDelight but doesn't own a cat or a rabbit.
5. The person in the white house has a cat.
6. David's house is blue and he doesn't buy BarkyMeal or Hop & Nibble.
7. Emma didn't buy AquaFresh or Whisk & Purr.
8. The dog's owner buys AquaFresh.
9. The rabbit owner thinks she likes Whisk & Purr because the yellow package matches her house.

Kakuro 8

	44↓	15↓	16↓	35↓	17↓	39↓		44↓	2↓	
36→							10→ 16↓			19↓
41→									1→ 7↓	
20→					29→ 15↓					
6→		24→ 3↓							2→ 1↓	
11→			18→ 3↓			9→ 9↓				
3→		36→ 20↓							3→	
15→			15↓	1↓	13→				7↓	■
21→					16↓	9↓	11→ 6↓			■
■	12→ 9↓			18→ 9↓				3→ 5↓		1↓
13→			27→						1→	

Hitori 8

2	5	6	4	1	7	8	3
4	3	5	1	3	6	8	7
7	6	3	7	2	5	2	4
5	3	5	4	7	1	8	2
2	7	4	6	2	8	5	8
3	4	3	5	2	2	1	8
7	8	1	2	5	4	3	4
5	1	6	3	4	2	7	5

Calcudoku 8

	3x	40x		
		15x		15x
	8x			
	3x		2	4
120x		10x		4x

Number Place 8

ACROSS

32, 55, 1457, 335, 56, 77, 426,
51, 65, 46656, 22636441, 146

DOWN

54, 5247, 635, 31, 466, 24,
1524447, 7761, 6335512, 51

Sudoku 8

			2				3	4
	9		8			6		
8				6	1	7		
1			3		5			
	4	9		1				
5	7	2						
	8						9	1
					2			
		1	4	9		3		

Number Search 8

```
4 9 9 2 4 2 5 5 4 8 0 8 8 6 3 3 7 2 1 0
5 7 9 8 7 7 2 5 2 7 9 2 0 2 9 8 4 3 5 2
5 4 1 0 7 5 7 9 1 0 5 8 4 2 9 0 1 3 9 1
2 7 0 0 4 0 1 8 2 7 5 5 6 6 0 0 7 6 4 5
5 1 0 6 9 9 8 4 6 4 7 0 0 5 8 4 5 8 1 4
1 1 0 3 6 1 2 6 9 6 0 3 0 0 8 8 3 4 9 4
7 2 3 1 6 9 9 2 1 6 8 9 9 1 4 2 9 4 3 4
9 0 7 8 7 5 4 2 2 1 1 4 7 2 7 8 0 8 8 7
3 6 6 5 5 8 2 8 6 7 1 9 3 6 5 4 8 6 5 7
1 7 9 1 9 4 9 4 3 6 6 9 0 9 3 2 6 5 0 2
9 7 2 1 1 2 6 3 7 4 5 0 2 2 1 8 2 4 6 8
3 1 6 0 5 1 7 5 1 4 3 1 3 6 8 5 8 2 4 4
1 4 7 5 1 9 8 6 3 2 2 3 2 6 1 3 3 3 0 0
7 8 1 2 1 2 9 7 5 6 7 5 6 9 3 7 7 3 7 9
7 1 2 9 9 4 8 7 3 9 5 8 9 8 9 9 4 2 9 4
8 3 7 3 4 0 0 3 6 9 7 0 8 5 1 2 0 7 6 3
8 7 3 7 0 7 5 8 9 3 4 9 9 6 7 6 1 8 4 3
6 2 8 2 8 3 4 9 3 3 6 5 3 4 2 0 6 7 8 3
6 4 7 1 8 0 0 5 8 0 9 9 5 1 9 3 5 3 6 5
5 2 1 4 9 7 5 8 2 9 3 5 2 9 4 9 1 0 3 7
```

- [ ] 965765
- [ ] 95908
- [ ] 9584219
- [ ] 087274
- [ ] 14961
- [ ] 4486542
- [ ] 176514
- [ ] 037692
- [ ] 190174
- [ ] 86111
- [ ] 39438
- [ ] 986582
- [ ] 917290
- [ ] 2942967
- [ ] 78372
- [ ] 273184
- [ ] 455242
- [ ] 333162
- [ ] 05842
- [ ] 18548

Chapter 5
Spicy

Number Place 9

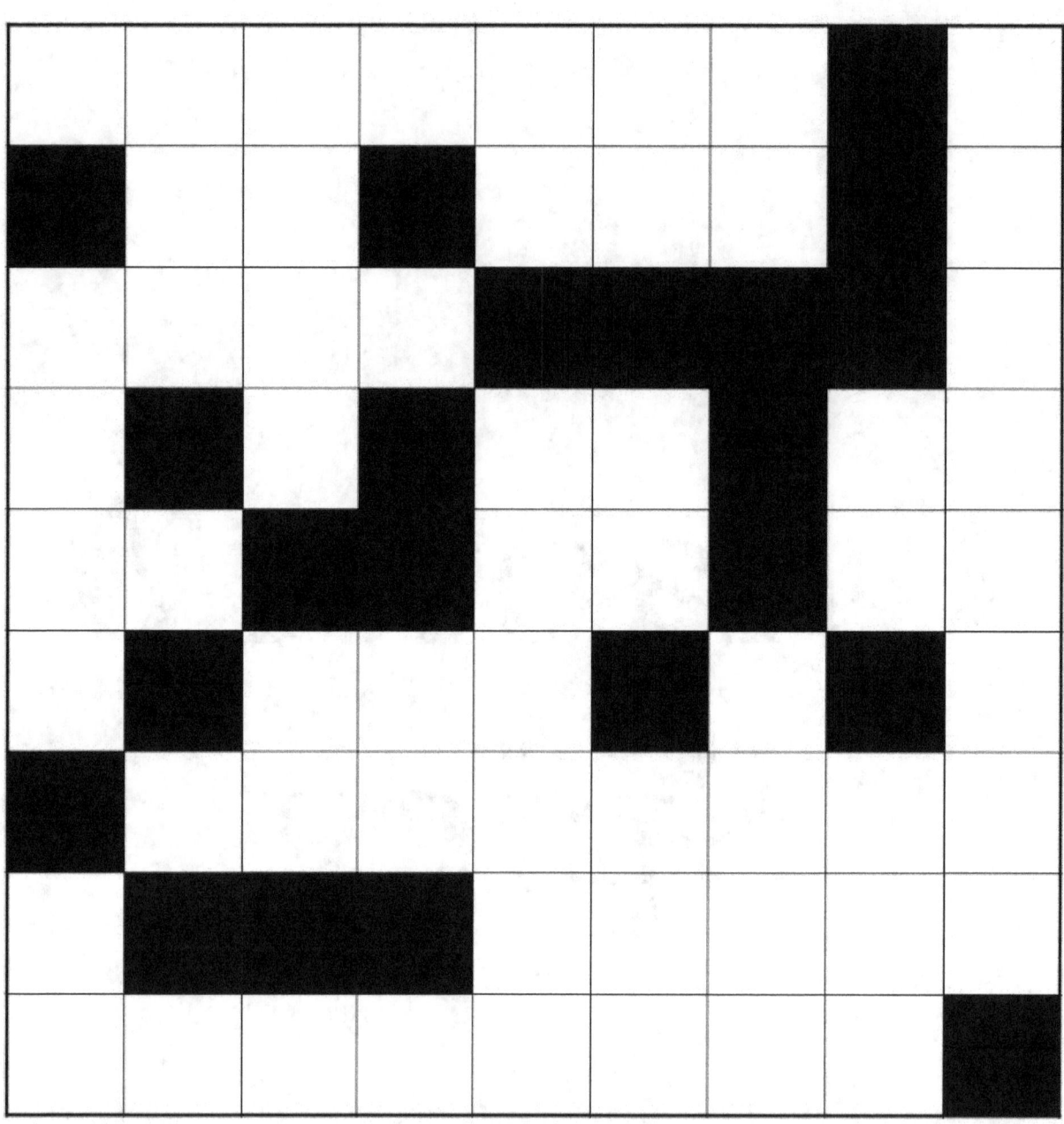

ACROSS

27, 578, 35, 7354, 485, 36, 71,
43247, 14247733, 48, 6888865,
62687215, 86

DOWN

8223, 76, 285847, 42, 8557, 143,
67, 58, 7734, 86, 74, 81, 833,
68118657, 737, 85

Sudoku 9

						4	8	1
		6		8	4	2	7	
						3	9	
5		2		9		7		
	3		8					2
	8		5	2				
	7	3		5				
1	6							
4	2					1	5	3

Hitori 9

1	6	3	2	8	7	5	4	3
2	8	5	6	2	6	7	9	4
9	5	2	4	8	1	3	8	7
7	1	6	2	4	4	8	5	2
4	4	9	8	5	3	6	5	1
5	9	1	3	7	9	4	8	2
6	2	4	9	4	8	7	3	5
8	2	7	5	6	5	9	2	3
4	7	8	2	3	5	2	6	9

Number Search 9

```
6 5 8 0 5 1 3 0 7 0 2 9 1 7 8 3 9 3 0 9 5 5 0 4
2 6 1 9 0 8 4 7 4 2 8 5 5 9 4 1 9 2 8 7 3 0 2 1
8 0 5 7 5 8 1 0 4 6 0 6 4 0 7 8 5 4 9 9 2 9 3 4
4 9 8 4 6 4 7 2 2 6 4 5 2 7 0 7 4 5 8 9 4 3 4 5
4 7 4 6 8 3 1 0 4 3 6 2 2 4 4 4 4 8 1 7 2 5 3 1
4 2 3 2 6 1 2 0 1 4 4 5 8 6 6 7 8 4 7 3 8 9 6 6
8 7 0 5 3 3 2 3 9 0 9 6 0 4 3 9 4 7 8 9 3 8 1 2
6 6 9 2 4 1 4 5 9 8 0 9 0 6 9 6 4 7 3 8 4 9 9 8
9 7 7 9 2 9 8 3 4 6 0 8 0 0 4 0 1 8 7 2 7 4 0 6
1 8 0 2 5 5 3 6 5 1 8 1 1 4 1 8 9 0 2 4 9 4 6 2
9 5 0 4 7 8 8 3 4 1 7 7 8 4 6 4 1 3 4 6 6 3 1 6
6 2 4 5 1 9 0 0 6 1 5 2 6 7 2 8 6 3 8 5 5 4 2 0
0 5 4 3 9 4 4 5 0 2 4 1 4 8 1 3 5 0 0 3 4 9 6 7
1 2 7 9 2 4 3 7 0 2 4 5 0 7 7 8 9 4 9 8 7 7 1 6
1 9 7 3 4 7 2 2 8 4 8 0 2 9 7 2 3 8 9 9 4 1 8 7
7 7 3 4 0 7 3 8 3 1 7 2 8 8 7 9 8 0 6 0 8 6 2 2
1 5 8 4 3 0 1 1 7 7 4 4 9 5 0 1 7 1 6 0 8 7 3 0
6 6 3 1 7 5 6 9 8 9 9 0 8 9 8 3 0 7 5 2 2 4 1 0
5 2 2 8 3 7 7 3 5 5 3 4 3 0 5 2 0 7 2 7 1 7 7 9
0 2 3 9 0 0 7 2 1 1 0 2 2 4 2 1 0 7 9 7 1 5 3 4
4 4 5 3 5 4 6 4 2 9 8 8 6 2 2 7 6 9 4 6 3 8 6 9
1 6 4 8 4 7 7 5 5 6 2 5 0 1 3 3 3 8 2 6 6 0 1 6
7 0 9 0 5 7 6 8 2 2 3 6 5 8 2 0 2 1 0 7 4 2 8 3
5 9 9 8 5 9 7 6 4 6 6 8 4 1 1 9 1 1 4 7 7 1 1 9
```

- ☐ 37132816
- ☐ 859974
- ☐ 70100814
- ☐ 01242201
- ☐ 9604394
- ☐ 54109
- ☐ 0014609
- ☐ 41432379
- ☐ 107966
- ☐ 41928730
- ☐ 3186414
- ☐ 3622444
- ☐ 521587
- ☐ 14056
- ☐ 08427
- ☐ 44218
- ☐ 23167
- ☐ 17518278
- ☐ 260381
- ☐ 008522

Logic Puzzle 9

		Flowers					Tool					Drinks				
		Roses	Tulips	Lilies	Sunflowers	Daffodils	Spade	Rake	Hoe	Shears	Watering Can	Lemonade	Iced Tea	Sparkling W	Soda	Fruit Punch
Friends	Sarah															
	Thomas															
	Ursula															
	Vincent															
	Wendy															
Drinks	Lemonade															
	Iced Tea															
	Sparkling W															
	Soda															
	Fruit Punch															
Tool	Spade															
	Rake															
	Hoe															
	Shears															
	Watering Can															

Garden Get-Together

Five friends – Sarah, Thomas, Ursula, Vincent, and Wendy – each planted a different type of flower in their gardens: roses, tulips, lilies, sunflowers, and daffodils. They also each have a favorite garden tool: spade, rake, hoe, shears, and watering can. When they get together, they each prefer a different drink: lemonade, iced tea, sparkling water, soda, and fruit punch.

Clues:

1. Ursula planted lilies and her favorite tool isn't the hoe.
2. Thomas drinks sparkling water and doesn't like the watering can or the rake.
3. The person who planted roses prefers the spade.
4. The person who drinks sparkling water prefers a hoe and the person who drinks fruit punch prefers a spade.
5. Sarah drinks iced tea and her favorite tool is the shears.
6. The person with the sunflowers doesn't like soda and prefers the hoe.
7. Vincent drinks fruit punch and didn't plant tulips or sunflowers.
8. Wendy's favorite tool is the watering can but she didn't plant daffodils.
9. The one who planted tulips prefers soda.

Calcudoku 9

1	2÷		1-		2-
	2÷			2-	
		2÷		4-	
360x		10x			1-
360x			6x		
		1	5	2÷	

Kakuro 9

	21↓	17↓	26↓	■	35↓	40↓	2↓	■	21↓	13↓	45↓	20↓	19↓	15↓
14→				7→ 6↓				31→ 16↓						
27→							42→							
13→				16→ 20↓			15→ 5↓			15→ 9↓				
31→								20→ 42↓					3→ 30↓	
■	45↓	30→ 14↓					1→ 25↓		8→ 21↓			3→ 9↓		■
7→		28→ 16↓								12→ 9↓				■
20→					44→									23↓
16→				11↓	36→ 18↓							12→ 12↓		
1→		18→ 7↓				18→ 19↓				19→ 12↓				
10→			31→ 8↓						9→ 6↓		8→ 22↓			
8→		2→ 15↓		9→ 20↓			21→ 14↓						9→	
24→					19→ 3↓					7→ 12↓		9↓	8↓	12↓
17→			13→				2↓	23→ 6↓						
9→			6→		30→						18→			

Logic Puzzle 10

	Instruments					Composer					Time Of Day				
	Violin	Trumpet	Piano	Flute	Guitar	Beethoven	Mozart	Chopin	Tchaikovsky	Bach	Morning	Afternoon	Evening	Late Night	Midday
Aaron															
Bella															
Charlie															
Donna															
Ethan															
Morning															
Afternoon															
Evening															
Late Night															
Midday															
Beethoven															
Mozart															
Chopin															
Tchaikovsky															
Bach															

Musical Maestros

Five friends – Aaron, Bella, Charlie, Donna, and Ethan – each play a different musical instrument: violin, trumpet, piano, flute, and guitar. Each has a favorite classical composer: Beethoven, Mozart, Chopin, Tchaikovsky, and Bach. When they practice, they each have a preferred time of day: morning, afternoon, evening, late night, and midday.

Clues:

1. Aaron plays the violin and doesn't prefer practicing in the evening.
2. Bella's favorite composer is Chopin, and she doesn't play the piano or the guitar, or in the afternoon.
3. The trumpet player practices in the late night and adores Mozart.
4. Charlie prefers practicing at midday but doesn't play the violin.
5. Donna adores Tchaikovsky but doesn't practice in the morning or afternoon.
6. The person who plays the guitar practices at midday and is a fan of Bach.
7. Ethan doesn't play the guitar or piano and his favorite composer isn't Beethoven.

Sudoku 10

5		4						
7		3				1		8
	6	8	3					
1			2		4			
2		6	8		7			4
4						8		
				4	5	2		3
			1				8	
		2		9	8	7	1	

Number Place 10

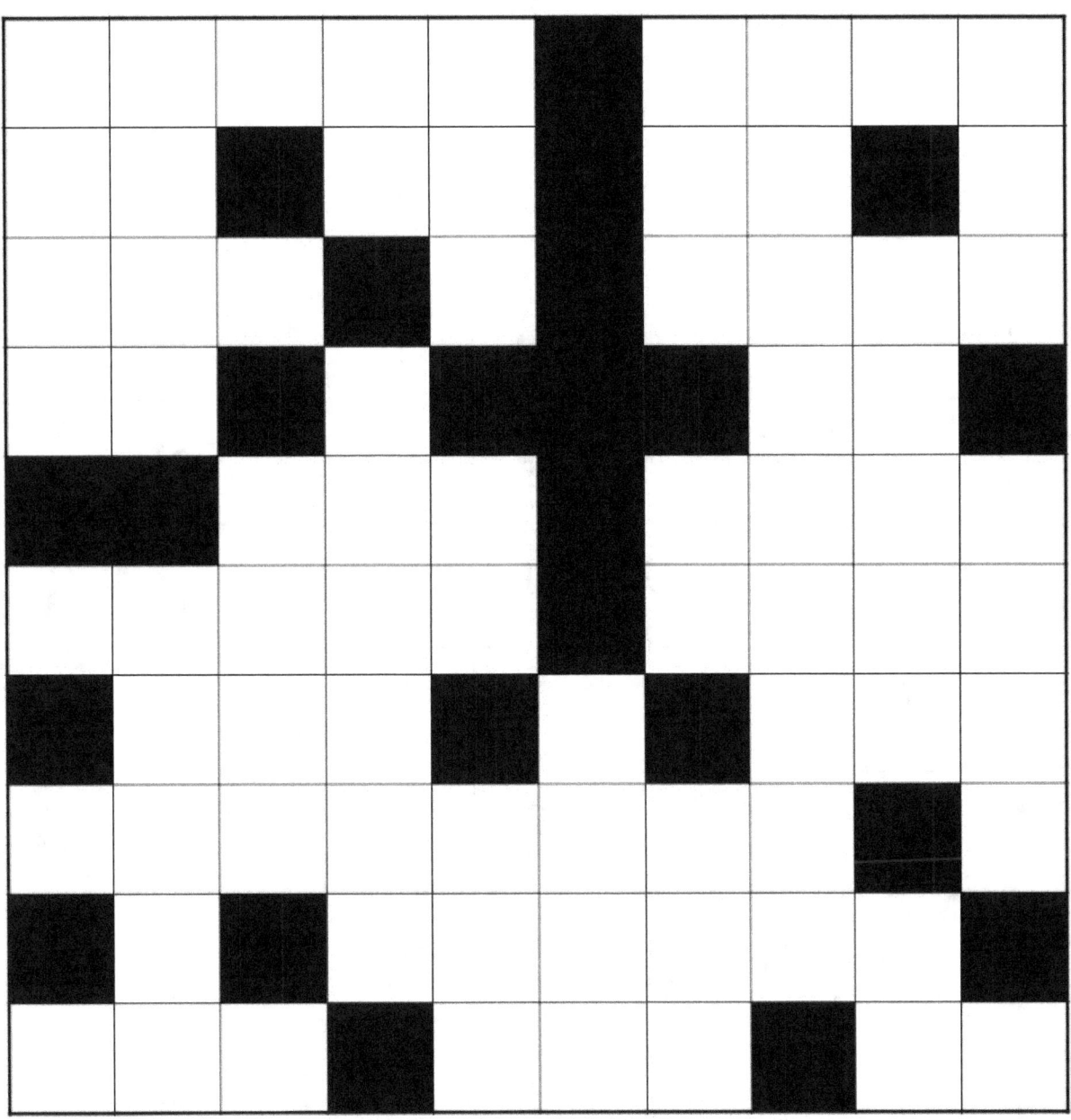

ACROSS

42, 837, 2379, 688, 98, 98226, 349, 969, 581, 233, 84414565, 36, 9533, 6737, 5589, 44, 75, 333791, 17673, 26

DOWN

9284, 368, 7835, 7391, 542, 5533, 12, 85438, 73, 432, 673, 1987, 523475359, 96, 989, 162113, 69, 74334

Hitori 10

4	1	6	4	2	2	8	5	9
1	3	1	8	9	6	1	4	9
9	4	3	5	4	1	7	8	2
4	2	8	7	4	1	5	3	1
5	9	7	1	8	2	8	6	4
7	8	3	4	3	4	6	2	5
7	6	9	3	1	5	6	7	3
1	7	4	9	5	8	2	3	2
3	7	2	6	8	4	9	1	8

Calcudoku 10

			4-	180x	
144x		4÷			
		7+		4	6
	5÷		2÷	2-	
		6			2
90x	2÷	270x			3-

Number Search 10

```
7 6 5 3 4 3 4 1 9 6 6 8 3 4 8 0 5 6 4 9 5 1 2 3
8 2 5 7 7 4 7 2 4 5 2 7 3 0 1 8 5 5 4 9 2 6 1 0
8 5 6 7 8 4 5 7 5 5 3 0 0 5 6 3 3 1 1 3 3 1 9 3
8 4 7 8 0 7 7 6 9 3 4 8 8 0 2 9 3 1 3 9 4 4 5 7
3 0 5 7 7 4 0 7 4 1 8 9 7 1 2 6 3 4 2 0 0 4 1 0
9 7 0 6 1 3 1 5 2 4 5 5 2 1 4 0 2 7 7 8 7 0 5 2
3 3 3 9 2 7 8 0 8 5 9 8 0 4 6 8 5 1 8 2 9 5 6 9
2 1 8 8 6 2 3 0 2 4 9 7 5 9 9 2 0 1 8 1 5 7 0 2
3 5 7 6 5 5 0 1 0 7 5 5 5 4 3 5 6 4 8 0 2 1 7 7
6 2 6 3 7 3 0 9 6 3 5 2 8 7 5 8 1 0 2 6 2 8 6 1
3 4 3 6 8 9 7 2 6 7 1 5 7 7 5 1 3 2 4 6 9 1 9 0
7 2 2 8 9 6 9 1 8 5 5 3 2 3 9 0 7 8 2 0 4 8 2 0
9 9 5 1 2 1 8 9 3 3 3 8 8 9 5 9 3 6 3 3 9 6 4 4
6 9 4 4 8 6 2 1 1 8 4 9 5 0 2 7 1 0 7 4 7 4 0 6
1 3 5 1 7 5 9 7 4 5 5 6 9 4 1 8 7 2 2 1 6 4 7 0
1 0 2 9 7 3 7 3 6 0 7 3 7 9 2 9 5 7 4 6 8 4 6 8
0 3 5 4 2 6 8 1 6 2 2 0 2 0 3 6 1 4 8 9 5 6 0 8
5 9 2 2 1 0 2 2 3 3 1 4 9 3 7 3 8 4 8 5 7 7 6 5
7 2 7 3 6 7 0 2 1 9 7 0 7 5 4 7 4 7 5 8 0 1 4 8
2 1 2 4 1 4 7 8 3 5 9 6 1 8 9 0 7 2 6 1 4 3 3 1
1 5 2 9 7 4 5 2 4 5 5 9 7 7 2 7 2 0 0 9 7 3 8 3
4 7 0 2 1 1 2 9 8 0 8 1 6 8 3 2 6 5 1 9 9 1 0 7
5 3 8 2 6 0 4 3 9 7 3 3 1 4 6 9 3 1 0 1 4 9 3 3
1 5 1 9 9 7 1 7 1 5 9 4 8 1 6 8 0 3 0 9 2 4 5 3
```

- [ ] 777537
- [ ] 69022654
- [ ] 62617
- [ ] 6857047
- [ ] 19173122
- [ ] 40247822
- [ ] 53434196
- [ ] 432491
- [ ] 82056900
- [ ] 452367
- [ ] 70020518
- [ ] 1277829
- [ ] 91964231
- [ ] 8739603
- [ ] 737454
- [ ] 515607
- [ ] 85980
- [ ] 49024339
- [ ] 323583
- [ ] 1013964

Kakuro 10

	19↓	45↓	■	29↓	20↓	7↓	29↓	9↓	28↓	5↓	11↓	23↓	■	
4→			45→ 21↓										45↓	12↓
37→								8→		18→ 24↓				
29→						2→ 2↓		10→ 45↓			14→ 20↓			
18→					40→ 3↓									3↓
■	18→ 13↓						24→ 37↓					7→ 12↓		
28→					19→ 11↓				20→ 19↓					13↓
12→			13↓	12→			14→				12→			
■	16→ 23↓			12→ 18↓			21→ 6↓				22→ 28↓			
15→					16→					4→ 26↓		11→ 2↓		
5→		22↓	6→ 3↓		5→		3→ 9↓		17→ 18↓					19↓
21→					26→ 13↓						■	2→ 9↓		
10→				9→ 12↓			29→ 11↓					13→ 5↓		
14→			14→			5→ 8↓		19→ 9↓					7→	
■	1→		34→								5→	■	6→	

Solutions

Number Search 1

```
9 1 0 3 8 0 3 9 7 4
7 9 9 7 3 2 6 3 4 6
5 0 1 3 5 1 1 0 2 9
3 1 1 9 2 3 1 1 2 9
6 0 1 5 6 1 7 5 8 0
9 0 9 5 5 9 8 1 3 2
7 5 3 5 1 8 5 6 8 4
9 1 4 5 8 8 3 9 4 3
3 6 5 5 7 4 9 4 6 5
7 5 3 9 5 0 7 8 8 4
```

Number Search 2

```
2 3 1 6 1 0 6 0 3 4
9 2 5 8 9 1 2 9 2 9
6 1 3 2 0 7 3 0 4 0
8 0 1 0 6 8 3 9 3 7
9 6 0 9 7 1 7 0 5 5
3 8 7 3 3 5 4 3 8 4
3 4 9 4 7 7 4 6 7
0 7 8 5 9 3 6 2 4 4
1 8 7 9 3 0 7 5 4 6
4 0 3 0 0 3 5 1 2 5
```

Number Search 3

```
8 7 8 1 6 0 1 0 0 5 3 9
0 8 5 4 6 7 3 3 9 4 1 2
4 7 5 7 3 5 5 5 5 2 4 0
8 9 2 7 1 5 7 8 6 8 7 0
7 0 9 4 8 9 2 8 2 2 6
2 9 4 5 0 3 0 6 3 1 6 6
8 5 3 5 4 3 9 9 8 4 5 2
6 0 8 6 4 7 1 3 8 7 9 2
2 9 6 7 0 5 9 5 0 1 6 7
8 5 3 9 0 0 4 5 1 2 4 9
```

Number Search 4

```
6 5 5 0 1 2 8 2 6 3 6 4
7 4 3 8 9 2 5 5 3 2 8 4
6 6 1 4 7 8 6 9 2 8 2 9
6 5 5 6 4 0 3 4 9 2 0 0
3 5 1 1 2 9 0 3 1 3 6 5
6 7 9 0 3 6 7 6 7 5 7 5
2 3 1 4 9 2 9 5 1 0 8 2
3 1 0 0 2 6 4 0 4 1 0 2
6 5 2 4 2 2 8 3 8 9 5 5
5 7 0 7 0 5 7 1 5 1 4 4
```

Number Search 5

```
9 3 8 2 1 2 8 1 3 6 0 0 2 2 3 6
7 7 9 2 1 8 2 1 9 2 5 8 6 8 1 1
7 7 6 8 2 7 2 8 5 0 8 5 8 6 6 9
3 1 4 0 7 2 1 4 3 9 9 4 5 9 7 7
1 8 5 4 5 5 6 4 6 0 2 3 9 3 1 4
0 5 9 7 3 7 7 1 9 5 3 7 8 8 9 2
7 0 9 8 4 9 5 5 7 4 8 1 6 2 8 5
6 7 1 1 5 4 8 8 5 0 2 4 2 0 7 9
6 9 4 3 4 0 7 9 4 4 7 0 4 4 0
4 7 2 9 0 3 5 6 3 9 8 4 6 7 4 1
3 7 7 9 8 5 9 7 4 4 6 6 4 9 3
3 7 6 6 6 4 4 3 7 2 1 4 3 5 6
5 6 3 8 1 0 9 4 9 5 8 6 7 6 3 8
2 4 0 7 1 9 0 7 5 1 3 4 8 3 6
1 4 0 6 5 3 9 2 3 2 7 6 9 2 0 6
6 9 2 3 3 2 6 4 0 0 6 9 9 1 8 1
```

Number Search 6

```
5 7 0 2 7 1 2 5 3 8 9 8 1 4 3 4
9 3 7 9 6 8 6 1 5 8 0 3 6 0 5 4
8 1 0 8 4 0 3 8 8 9 2 4 5 3 9 4
5 6 3 1 9 9 5 3 9 6 4 2 7 9 6 6
9 9 6 4 5 7 5 0 2 6 0 7 1 4 5 3
5 1 0 3 8 0 6 0 6 4 9 5 2 5 8 6
8 5 9 0 5 5 7 2 5 4 1 3 6 3 4 3
9 2 5 8 9 0 4 9 8 1 8 9 0 9 3 5
8 6 0 7 9 6 7 0 3 2 9 3 5 4 1 1
1 6 2 2 5 2 6 2 9 9 1 5 9 9 0 1 9
9 5 4 0 9 7 5 6 8 4 0 4 0 1 1 0
6 8 4 3 5 3 7 0 3 1 7 9 7 3 4 8 0
0 8 3 7 8 5 9 8 3 2 1 4 7 1 2 6
1 8 3 9 4 9 9 5 3 0 5 8 9 4 0 0
2 6 2 3 2 8 5 8 7 5 3 7 7 9 5 0
2 6 7 7 5 7 0 7 6 1 5 3 5 9 6 6
```

Number Search 7

```
6 4 4 0 5 3 0 5 9 5 0 4 0 0 9 8 3 0 7 1
5 3 2 2 6 1 5 5 5 5 5 0 1 9 4 7 4 5 6 7
6 7 1 1 4 6 3 7 1 6 4 0 4 5 9 3 2 0 5 4
9 5 1 6 2 9 0 9 2 9 1 9 3 3 8 9 7 8 1 1
2 8 1 6 3 9 1 2 6 4 2 2 2 8 7 6 5 2 8 6
7 4 2 0 6 0 0 1 7 5 1 2 3 1 5 6 9 8 9 8
6 8 7 4 5 0 0 7 5 0 0 8 1 6 5 1 8 2 0 9
8 4 4 2 8 7 5 9 7 0 9 9 8 6 8 3 8 1 1 0
6 2 0 2 4 5 2 3 0 9 0 3 8 3 5 2 0 5 3 4
1 7 3 6 0 4 8 1 7 8 1 9 8 5 9 6 6 1 0 9
8 4 8 5 9 0 4 1 2 6 4 7 7 9 2 4 5 6 2 3
7 7 3 9 3 9 0 4 4 0 5 8 9 0 9 3 7 2 0 2
1 9 1 3 8 7 8 7 2 7 6 1 2 3 6 5 3 3 9 6
6 7 9 4 9 4 9 9 9 0 6 1 9 2 2 4 9 4 6 0
4 0 8 7 1 2 8 0 4 0 5 8 9 1 5 7 8 8 2 5
1 4 9 6 9 7 1 2 5 2 8 0 9 1 6 3 4 6 4 5
3 5 6 3 7 8 4 1 0 3 2 6 2 7 0 3 1 8 8
8 6 3 2 2 5 2 7 3 7 7 8 9 3 8 7 5 0 4 8
3 2 7 2 2 6 8 3 2 1 8 0 7 9 0 1 1 0 3 2
9 1 7 2 9 4 6 2 4 5 4 9 2 9 5 5 5 0 6 6
```

Number Search 8

```
4 9 9 2 4 2 5 5 4 8 0 8 8 6 3 3 7 2 1 0
5 7 9 8 7 7 2 5 2 7 9 2 0 2 9 8 4 3 5 2
5 4 1 0 7 5 7 9 1 0 5 8 4 2 9 0 1 3 9 1
2 7 0 4 0 1 8 2 7 5 5 6 6 0 0 7 6 4 5
5 1 0 6 9 9 8 4 6 4 7 0 0 5 8 4 5 8 1 4
1 1 0 3 6 1 2 6 9 6 0 3 0 0 8 3 4 9 4
7 2 3 1 6 9 2 1 6 9 9 1 4 2 9 4 3 4
9 0 7 8 7 5 4 2 2 1 1 4 7 2 7 8 0 8 8 7
3 6 6 5 5 8 2 8 6 7 1 9 3 6 5 4 8 6 5 7
1 7 7 1 9 4 3 6 6 9 0 9 3 2 6 5 0 2
9 7 2 1 1 2 6 3 7 4 5 0 2 2 1 8 2 4 6 8
3 1 6 0 5 1 7 5 1 4 3 1 3 6 8 5 8 2 4 4
1 4 7 5 1 9 8 6 3 2 2 3 2 6 1 3 3 0 0
7 8 1 2 1 2 9 7 5 6 7 5 6 9 3 7 7 3 7 9
7 1 2 9 4 8 7 3 9 8 9 8 9 4 2 4
8 3 7 3 4 0 0 3 6 9 7 0 8 5 1 2 0 7 6 3
8 7 3 7 0 7 5 8 9 3 4 9 9 6 7 6 1 8 4 3
6 2 8 2 8 3 4 9 3 3 6 5 3 4 2 0 6 7 8 3
6 4 7 1 8 0 0 5 8 0 9 9 5 1 9 3 5 3 6 5
5 2 1 4 9 7 5 8 2 9 3 5 2 9 4 9 1 0 3 7
```

Number Search 9

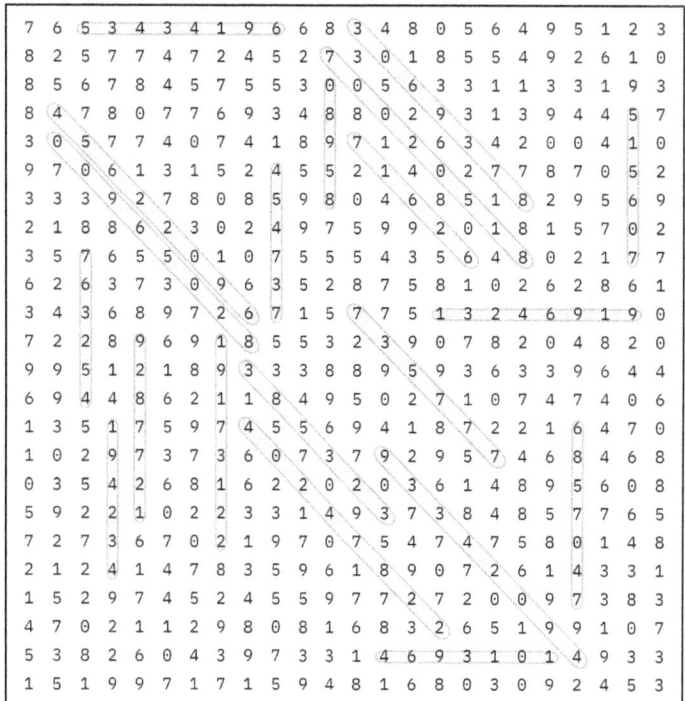

```
6 5 8 0 5 1 3 0 7 0 2 9 1 7 8 3 9 3 0 9 5 5 5 0 4
2 6 1 9 0 8 4 7 4 2 8 5 5 9 4 1 9 2 8 7 3 0 2 1
8 0 5 7 5 8 1 0 4 6 0 6 4 0 7 8 5 4 9 9 2 9 3 4
4 9 8 4 6 4 7 2 2 6 4 5 2 7 0 7 4 5 8 9 4 3 4 5
4 7 4 6 8 3 1 0 4 3 6 2 2 4 4 4 8 1 7 2 5 3 1
4 2 3 2 6 1 2 0 1 4 4 5 8 6 6 7 8 4 7 3 8 9 6 6
8 7 5 0 3 3 2 3 9 0 9 6 0 4 3 9 4 7 8 9 3 8 1 2
6 6 9 2 4 1 4 5 9 8 0 9 0 6 9 6 4 7 3 8 4 9 9 8
9 7 7 9 2 9 8 3 4 6 0 8 0 0 4 0 1 8 7 2 7 4 0 6
1 8 0 2 5 5 3 6 5 1 8 1 1 4 1 8 9 0 2 4 9 4 6 2
9 5 0 4 7 8 8 3 4 1 7 7 8 4 6 4 1 3 4 6 6 3 1 6
6 2 4 5 1 9 0 0 6 1 5 2 6 7 2 8 6 3 8 5 5 4 2 0
0 5 4 3 9 4 4 5 0 2 4 1 4 8 1 3 5 0 0 3 4 9 6 7
1 2 7 9 2 4 3 7 0 2 4 5 0 7 7 8 9 4 9 8 7 7 1 6
1 9 7 3 4 7 2 2 8 4 8 0 2 9 7 2 3 8 9 9 4 1 8 7
7 7 3 4 0 7 3 8 3 1 7 2 8 8 7 9 8 0 6 0 8 6 2 2
1 5 8 4 3 0 1 1 7 4 4 9 5 0 1 7 1 6 0 8 7 3 0
6 6 3 1 7 5 6 9 8 9 9 0 8 9 8 3 0 7 5 2 2 4 1 0
5 2 2 8 3 7 7 3 5 5 3 4 3 0 5 2 0 7 2 7 1 7 7 9
0 2 3 9 0 0 7 2 1 1 0 2 2 4 2 1 0 7 9 7 1 5 3 4
4 4 5 3 5 4 6 4 2 9 8 8 6 2 2 7 6 9 4 6 3 8 6 9
1 6 4 8 4 7 7 5 5 6 2 5 0 1 3 3 3 8 2 6 6 0 1 6
7 0 9 0 5 7 6 8 2 2 3 6 5 8 2 0 2 1 0 7 4 2 8 3
5 9 9 8 5 9 7 9 6 4 6 6 6 8 4 1 1 9 1 1 4 7 7 1 1 9
```

Number Search 10

```
7 6 5 3 4 3 4 1 9 6 6 8 3 4 8 0 5 6 4 9 5 1 2 3
8 2 5 7 7 4 7 2 4 5 2 7 3 0 1 8 5 5 4 9 2 6 1 0
8 5 6 7 8 4 5 7 5 5 3 0 0 5 3 3 1 1 3 3 1 9 3
8 4 7 8 0 7 7 6 9 3 4 8 8 0 2 9 3 1 3 9 4 4 5 7
3 0 5 7 7 4 0 7 4 1 8 9 7 1 2 6 3 4 2 0 0 4 1 0
9 7 0 6 1 3 1 5 2 4 5 5 2 1 4 0 2 7 7 8 7 0 5 2
3 3 3 9 2 7 8 0 8 5 9 8 0 4 6 8 5 1 8 2 9 5 6 9
2 1 8 8 6 2 3 0 2 4 9 7 5 9 9 2 0 1 8 1 5 7 0 2
3 5 7 6 5 5 0 1 0 7 5 5 5 4 3 5 6 4 8 0 2 1 7 7
6 2 6 3 7 3 0 9 6 3 5 2 8 7 5 8 1 0 2 6 2 8 6 1
3 4 3 6 8 9 7 2 6 7 1 5 7 7 5 1 3 2 4 6 9 1 9 0
7 2 2 8 2 2 6 9 1 8 5 5 3 2 3 9 0 7 8 2 0 4 8 2 0
9 9 5 1 2 1 8 9 3 3 3 8 8 9 5 9 3 6 3 3 9 6 4 4
6 9 4 4 8 6 2 1 1 8 4 9 5 0 2 7 1 0 7 4 7 4 0 6
1 3 5 1 7 5 9 7 4 5 5 6 9 4 1 8 7 2 2 1 6 4 7 0
1 0 2 9 7 3 7 3 0 7 3 7 9 2 9 5 7 4 6 8 4 6 8
0 3 5 4 2 6 2 8 1 6 2 2 0 2 0 3 6 1 4 8 9 5 6 0 8
5 9 2 2 1 0 2 3 0 3 1 4 9 3 7 3 8 4 8 5 7 7 6 5
7 2 7 3 6 7 0 2 1 9 7 0 7 5 4 7 4 7 5 8 0 1 4 8
2 1 2 4 1 4 7 8 3 5 9 6 1 8 9 0 7 2 6 1 4 3 3 1
1 5 2 9 7 4 5 2 4 5 5 9 7 7 2 7 2 0 0 9 7 3 8 3
4 7 0 2 1 1 2 9 8 0 8 1 6 8 3 2 6 5 1 9 9 1 0 7
5 3 8 2 6 0 4 3 9 7 3 3 1 4 6 9 3 1 0 1 4 9 3 3
1 5 1 9 9 7 1 7 1 5 9 4 8 1 6 8 0 3 0 9 2 4 5 3
```

97

Kakuro 1

Kakuro 2

Kakuro 3

Kakuro 4

Kakuro 5

Kakuro 6

Kakuro 7

	38↓	20↓	36↓	35↓			12↓	13↓	5↓	10↓
20→	6	9	3	2	25↓	16→ 45↓	3	2	5	6
42→	2	5	1	8	7	4	9	6	4→ 30↓	4
29→	4	6	7	3	1	8	8→	5	3	7↓
9→	9	19→ 23↓	2	6	8	3	10↓	15→	8	7
37→	1	4	6	7	9	2	8	5→ 14↓	5	31↓
19→	5	2	8	4	28→ 18↓	6	2	8	7	5
36→	8	1	9	5	6	7	16→ 23↓	6	1	9
10→	3	7	13↓	20→ 5↓	7	5	8	11→	4	7
	35→ 7↓	3	8	4	5	9	6	8→ 2↓	2	6
19→	7	6	5	1	12→	1	9	2	4→	4

Kakuro 8

	44↓	15↓	16↓	35↓	17↓	39↓		44↓	2↓	
36→	4	3	7	8	9	5	10→ 16↓	8	2	19↓
41→	5	7	2	9	8	1	3	6	1→ 7↓	1
20→	2	5	6	7	29→ 15↓	3	9	2	7	8
6→	6	24→ 3↓	1	2	3	9	4	5	2→ 1↓	2
11→	8	3	18→ 3↓	5	7	6	9→ 9↓	3	1	5
3→	3	36→ 20↓	3	4	5	8	7	9	3→	3
15→	7	8	15↓	1	13→	7	2	4	7↓	
21→	9	3	8	1	16↓	9	11→ 6↓	7	4	
	12→ 9↓	5	7	18→ 9↓	9	7	2	3→ 5↓	3	1↓
13→	9	4	27→	9	7	2	4	5	1→	1

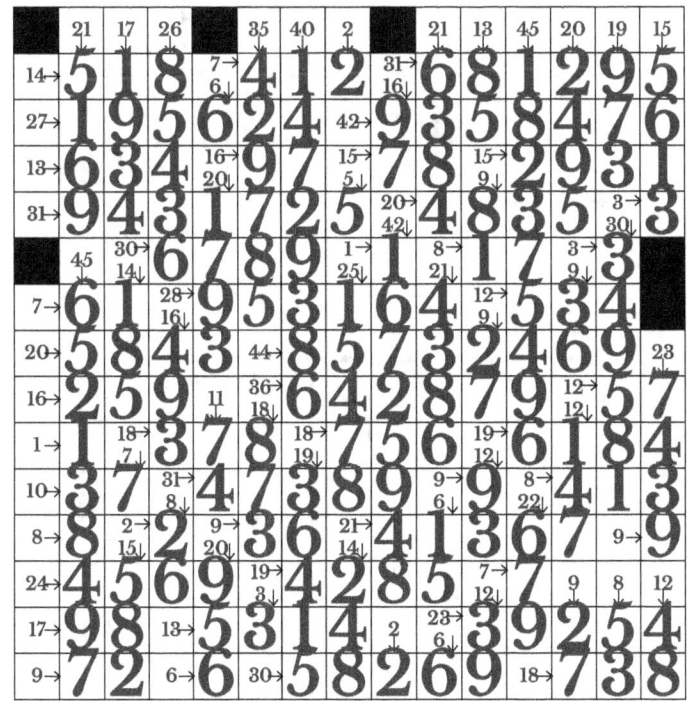

Kakuro 9

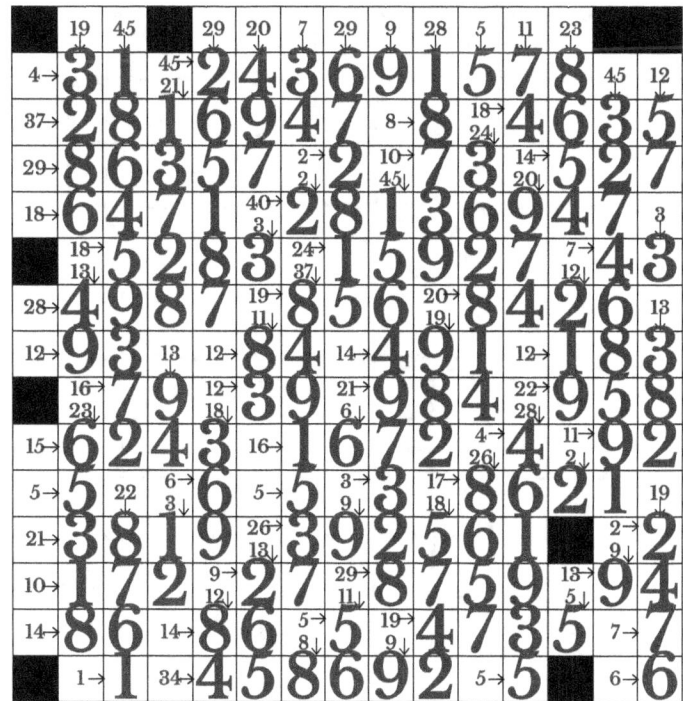

Kakuro 10

99

Sudoku 1

3	5	2	7	6	1	9	8	4
1	6	7	4	8	9	3	5	2
4	8	9	2	5	3	1	6	7
8	3	6	1	9	2	4	7	5
2	4	1	5	7	6	8	3	9
7	9	5	3	4	8	6	2	1
5	7	8	6	1	4	2	9	3
6	1	3	9	2	7	5	4	8
9	2	4	8	3	5	7	1	6

Sudoku 2

1	7	4	8	9	5	3	6	2
5	2	3	7	4	6	1	9	8
8	9	6	1	3	2	5	4	7
7	5	8	4	2	9	6	1	3
6	1	9	3	7	8	4	2	5
3	4	2	5	6	1	7	8	9
9	3	1	6	8	7	2	5	4
2	6	7	9	5	4	8	3	1
4	8	5	2	1	3	9	7	6

Sudoku 3

9	3	7	5	2	4	8	6	1
8	5	6	1	7	9	2	3	4
4	1	2	6	3	8	7	9	5
2	7	8	3	6	1	4	5	9
3	9	4	8	5	7	6	1	2
1	6	5	9	4	2	3	8	7
5	4	1	2	8	6	9	7	3
6	2	9	7	1	3	5	4	8
7	8	3	4	9	5	1	2	6

Sudoku 4

7	1	3	2	9	5	6	4	8
4	6	8	7	3	1	2	9	5
2	9	5	4	6	8	1	3	7
1	8	9	6	2	4	5	7	3
6	5	4	3	1	7	9	8	2
3	7	2	8	5	9	4	1	6
8	4	1	5	7	2	3	6	9
9	2	6	1	8	3	7	5	4
5	3	7	9	4	6	8	2	1

Sudoku 5

2	6	4	5	9	3	1	8	7
7	9	8	6	2	1	4	3	5
5	3	1	7	4	8	2	9	6
3	5	7	2	1	4	8	6	9
6	1	9	3	8	7	5	2	4
4	8	2	9	5	6	7	1	3
1	4	3	8	7	9	6	5	2
8	2	6	4	3	5	9	7	1
9	7	5	1	6	2	3	4	8

Sudoku 6

3	4	5	1	7	6	9	8	2
6	1	7	9	8	2	5	3	4
2	9	8	3	5	4	1	6	7
9	2	4	8	3	5	6	7	1
7	8	6	2	1	9	4	5	3
1	5	3	4	6	7	8	2	9
8	6	1	7	4	3	2	9	5
5	3	9	6	2	1	7	4	8
4	7	2	5	9	8	3	1	6

Sudoku 7

8	7	3	5	2	6	9	4	1
6	1	4	3	7	9	5	2	8
5	9	2	8	4	1	3	6	7
2	5	7	9	8	3	6	1	4
3	6	8	1	5	4	7	9	2
9	4	1	2	6	7	8	3	5
4	3	6	7	1	5	2	8	9
7	2	9	4	3	8	1	5	6
1	8	5	6	9	2	4	7	3

Sudoku 8

6	1	5	2	7	9	8	3	4
2	9	7	8	3	4	6	1	5
8	3	4	5	6	1	7	2	9
1	6	8	3	2	5	9	4	7
3	4	9	6	1	7	5	8	2
5	7	2	9	4	8	1	6	3
4	8	6	7	5	3	2	9	1
9	5	3	1	8	2	4	7	6
7	2	1	4	9	6	3	5	8

Sudoku 9

2	9	7	3	6	5	4	8	1
3	1	6	9	8	4	2	7	5
8	5	4	7	1	2	3	9	6
5	4	2	1	9	6	7	3	8
6	3	9	8	4	7	5	1	2
7	8	1	5	2	3	6	4	9
9	7	3	2	5	1	8	6	4
1	6	5	4	3	8	9	2	7
4	2	8	6	7	9	1	5	3

Sudoku 10

5	1	4	9	8	2	3	7	6
7	2	3	4	5	6	1	9	8
9	6	8	3	7	1	5	4	2
1	8	9	2	3	4	6	5	7
2	5	6	8	1	7	9	3	4
4	3	7	5	6	9	8	2	1
8	9	1	7	4	5	2	6	3
6	7	5	1	2	3	4	8	9
3	4	2	6	9	8	7	1	5

Logic Puzzle 1

Friends	Meal	Cars	Event
Allan	Lasagna	Ford	Baby
Brenda	Pie	Toyota	Grad
Craig	Salad	Nissan	Promo
Donna	Sushi	Honda	House
Ella	Chick	Tesla	Engage

Logic Puzzle 2

nds	Cabin Color	Activity	Beverage
Marco	Red	Hike	Coffee
Nina	Green	Bird	Hot
Oli	White	Read	Lemon
Penny	Yellow	Swim	Water
Quinn	Blue	Paint	Tea

Logic Puzzle 3

Neighbors	Pets	House Color	Fruits
Sarah	Dog	White	Elder
Tim	Fish	Pink	Apple
Ursula	Parrot	Blue	Cherry
Victor	Cat	Green	Date
Wendy	Rabbit	Brown	Banana

Logic Puzzle 4

Neighbors	Flowers	Tool	Season
Zack	Sun	Shears	Winter
Yara	Daisies	Rake	Fall
Xavier	Tulips	Hoe	Rainy
Yvonne	Rose	Spade	Sum
Zoe	Lilies	Trowel	Spring

Logic Puzzle 5

Friends	Band	Shirt Color	Snack
Amy	Lunar	White	Ice Cre
Brian	Quant	Black	Nachos
Carla	Red Val	Green	Popco
Derek	Melody	Yellow	Pretzel
Evan	Sonic	Blue	Fruit

Logic Puzzle 6

Friends	Book	Bag	Beverage
Fiona	Oceans	Mess	Water
Greg	Digital	Tote	Soda
Helen	Magic	Back	Coffee
Ian	An Civ	Sling	Ho Cho
Julie	Myster	Hand	Tea

Logic Puzzle 7

Friends	Exhibit	Transport	Souvenir
Lauren	Abstr	Bike	Note
Mark	Real	Bus	Scarf
Nancy	Sculpt	Car	Key
Oscar	Impress	Subway	Poster
Penny	Digital	Walk	Mug

Logic Puzzle 8

Friends	Pets	House Color	Pet Food
Alice	Rabbit	Yellow	Whisk
Bob	Fish	Green	Barky
Claire	Parrot	Red	Feather
David	Dog	Blue	Aqua
Emma	Cat	White	Hop

Logic Puzzle 9

Friends	Flowers	Tool	Drinks
Sarah	Daffo	Shears	Tea
Thomas	Sun	Hoe	Spark
Ursula	Lilies	Rake	Lemon
Vincent	Roses	Spade	Punch
Wendy	Tulips	Can	Soda

Logic Puzzle 10

Friends	Instruments	Composer	Time Of Day
Aaron	Violin	Beetho	Afterno
Bella	Flute	Chopin	Morn
Charlie	Guitar	Bach	Mid
Donna	Piano	Tchaik	Even
Ethan	Trump	Mozart	Late

Calcudoku 1

⁷⁺2	4	1	⁸3
1	2	³3	⁷⁺4
4	3	2	⁶⁻1
⁸⁺3	⁶⁺1	4	2

Calcudoku 2

⁴4	⁶⁺1	2	3
³⁺1	2	3	4
⁵⁺2	³3	⁷⁺4	1
3	⁵⁺4	1	⁷⁺2

Calcudoku 3

⁶⁺1	4	⁵⁺3	2
⁹⁻3	1	²2	⁸⁻4
4	2	¹1	3
2	3	⁹⁺4	1

Calcudoku 4

⁴4	1	⁶⁺5	³⁺2	3
⁷⁺2	5	3	1	⁷⁺4
3	2	⁷⁺4	⁶⁺5	1
5	¹⁴⁺4	1	3	⁸⁺2
1	⁴⁺3	2	⁴4	⁵5

Calcudoku 5

4	5	³3	1	¹⁻2
1	¹⁴⁺4	¹³⁺2	3	5
2	¹1	5	⁴4	3
5	3	4	2	1
¹⁰⁺3	¹⁴⁺2	⁴⁻1	5	¹⁻4

Calcudoku 6

¹⁻2	1	¹⁻4	3	⁵5
1	²2	5	¹⁻4	3
4	3	²2	³⁻5	1
²⁻3	5	⁴⁻1	2	4
5	⁴4	3	²⁻1	¹⁰⁺2

Calcudoku 7

3	5	2	1	¹²ˣ4
2	1	⁵ˣ5	⁸ˣ4	3
¹⁵⁰ˣ5	²2	⁴4	³ˣ3	1
¹²ˣ1	4	³⁰ˣ3	2	¹⁰ˣ5
⁴4	3	1	5	2

Calcudoku 8

1	³ˣ3	⁴⁰ˣ4	5	2
2	4	¹⁵ˣ1	3	¹⁵ˣ5
⁸ˣ4	2	5	1	3
5	³ˣ1	3	²2	⁴4
¹²⁰ˣ3	5	¹⁰ˣ2	4	⁴ˣ1

Calcudoku 9

¹1	²÷2	4	¹⁻6	3	²⁻5
2	²÷1	6	4	²⁻5	3
4	5	²÷3	1	⁴2	6
³⁶⁰ˣ6	3	¹⁰ˣ5	2	1	¹⁻4
³⁶⁰ˣ5	4	2	⁶ˣ3	6	1
3	6	¹1	⁵5	²÷4	2

Calcudoku 10

2	3	4	⁴⁻1	¹⁸⁰ˣ6	5
¹⁴⁺ˣ4	6	⁴⁺1	5	2	3
3	1	⁷⁺5	2	⁴4	6
6	⁵⁻5	2	²⁺4	3	1
1	4	⁶6	3	5	²2
⁹⁰ˣ5	²⁺2	²⁷⁰ˣ3	6	1	³⁻4

103

Number Place 1

1	■	■	2
■	1	■	1
■	2	2	2
1	2	2	3

Number Place 2

2	3	2	3
2	2	1	3
3	3	1	■
■	■	■	■

Number Place 3

3	■	■	■	3
■	4	2	■	4
■	■	■	■	4
4	1	1	2	■
4	4	1	2	1

Number Place 4

2	4	1	2	■
2	2	■	3	4
2	■	2	■	2
■	3	■	■	3
■	■	■	1	2

Number Place 5

1	2	1	2	3	4
1	1	■	■	■	■
2	3	3	5	■	2
4	■	1	2	■	■
2	■	■	■	1	4
5	5	5	5	4	■

Number Place 6

5	6	1	2	5	■	5
■	■	6	1	■	■	1
4	■	■	6	2	4	1
6	5	2	■	3	3	■
■	3	4	2	1	2	6
1	1	3	■	3	■	4
3	6	1	■	6	■	5

Number Place 7

7	7	3	■	■	■	6	2
■	■	■	■	1	1	2	4
6	2	1	2	5	■	4	1
■	2	6	2	■	■	2	■
■	■	4	2	6	1	1	7
1	4	■	5	6	3	■	5
5	7	7	1	7	■	3	5
4	2	■	7	7	■	3	■

Number Place 8

4	6	6	5	6	■	4	■
■	3	■	4	■	1	■	■
3	3	5	■	■	5	6	■
■	5	1	■	3	2	■	7
5	5	■	■	1	4	5	7
■	1	4	6	■	4	2	6
2	2	6	3	6	4	4	1
4	■	6	5	■	7	7	■

Number Place 9

6	8	8	8	8	6	5	■	6
■	3	5	■	5	7	8	■	8
7	3	5	4	■	■	■	■	1
7	■	7	■	2	7	■	7	1
3	6	■	■	8	6	■	4	8
4	■	4	8	5	■	8	■	6
■	6	2	6	8	7	2	1	5
8	■	■	■	4	3	2	4	7
1	4	2	4	7	7	3	3	■

Number Place 10

1	7	6	7	3	■	5	5	8	9
9	8	■	3	6	■	4	2	■	8
8	3	7	■	8	■	2	3	7	9
7	5	■	1	■	■	■	4	4	■
■	■	9	6	9	■	6	7	3	7
9	8	2	2	6	■	9	5	3	3
■	5	8	1	■	5	■	3	4	9
8	4	4	1	4	5	6	5	■	1
■	3	■	3	3	3	7	9	1	■
6	8	8	■	2	3	3	■	2	6

Hitori 1

1	2	**[2]**	4
[1]	4	1	3
3	**[2]**	4	2
4	3	2	**[2]**

Hitori 2

[4]	1	**[1]**	4
2	4	1	3
[3]	3	4	2
4	2	3	1

Hitori 3

2	5	4	**[2]**	3
1	**[3]**	3	2	5
4	1	5	**[2]**	2
5	**[4]**	2	4	1
3	2	1	5	4

Hitori 4

1	2	4	5	3
5	3	1	**[3]**	2
2	5	3	1	**[1]**
4	1	**[5]**	3	5
3	4	5	**[1]**	1

Hitori 5

[3]	3	5	2	**[2]**	1
6	1	**[5]**	5	2	3
1	6	2	4	3	5
[4]	2	**[5]**	6	**[2]**	4
2	5	1	3	4	**[1]**
[1]	4	6	**[3]**	5	2

Hitori 6

5	7	4	**[1]**	3	6	1
3	**[5]**	1	6	**[2]**	2	7
[3]	6	2	**[5]**	1	7	**[7]**
7	2	5	1	6	**[6]**	4
[5]	1	**[1]**	4	3	5	3
6	5	3	7	4	1	2
1	**[2]**	7	5	**[4]**	4	6

Hitori 7

8	**[6]**	4	**[1]**	6	3	**[1]**	1
6	4	1	7	8	2	3	5
5	1	**[4]**	4	3	**[3]**	8	**[2]**
4	**[1]**	8	**[6]**	7	1	**[6]**	6
2	7	3	**[6]**	8	8	1	4
[5]	6	**[6]**	8	1	**[1]**	4	2
7	8	6	1	**[3]**	5	2	3
1	**[8]**	7	3	2	4	6	8

Hitori 8

2	5	6	**[4]**	1	7	**[8]**	3
4	**[3]**	5	1	3	6	8	7
[7]	6	3	7	**[2]**	5	2	4
5	3	**[5]**	4	7	1	**[8]**	2
[2]	7	4	6	2	**[8]**	5	**[8]**
3	4	**[3]**	5	**[2]**	2	1	8
7	8	1	2	5	4	3	**[4]**
[5]	1	**[6]**	3	4	**[2]**	7	5

Hitori 9

1	6	3	2	**[8]**	7	5	4	**[3]**
[2]	8	5	**[6]**	2	6	**[7]**	9	4
9	5	2	4	8	1	3	**[8]**	7
7	**[1]**	6	**[2]**	4	**[4]**	8	5	2
[4]	4	9	8	5	3	6	**[5]**	1
5	9	1	3	7	**[9]**	4	8	**[2]**
6	2	4	9	**[4]**	8	7	3	5
8	**[2]**	7	5	6	**[5]**	9	2	3
4	7	8	**[2]**	3	5	2	**[6]**	9

Hitori 10

4	1	6	**[4]**	2	**[2]**	8	5	9
[1]	3	**[1]**	8	9	6	1	4	**[9]**
9	4	3	5	**[4]**	1	7	8	2
[4]	2	8	7	4	**[1]**	5	**[3]**	1
5	9	7	1	8	2	**[8]**	6	4
7	8	**[3]**	4	3	**[4]**	6	2	5
[7]	6	9	**[3]**	1	5	**[6]**	7	3
1	7	4	9	5	8	2	3	**[2]**
3	**[7]**	2	6	8	**[4]**	9	1	8

We create all of the puzzles in this book, so if you run into any trouble or find a mistake, we would love to hear about it so we can fix it in future editions. Email sally@sportsmomsarethebest.com with suggestions for changes that would make the puzzles better.

We are also open to future product suggestions. If you would love to see a book of your favorite puzzle that you have a hard time finding in full books, please let us know.

Thanks!
Elizabeth and the team at Sports Moms are the Best Press